Looking at Stars

Planets

Robin Kerrod

Thameside Press

U.S Publication copyright © 2001 Thameside Press.
International copyright reserved in all countries.
No part of this book may be reproduced in any
form without written permission from the publisher.

Distributed in the United States by
Smart Apple Media
1980 Lookout Drive
North Mankato, MN 56003

Text copyright © by Robin Kerrod 2001

Editor: Veronica Ross
Designer: Helen James
Illustrator: Chris Forsey
Consultant: Doug Millard
Picture researcher: Diana Morris

Printed in Hong Kong

Library of Congress Cataloging-in-Publication Data

Kerrod, Robin.
   Planets / written by Robin Kerrod.
     p. cm. -- (Looking at stars)
   Includes index.
   ISBN 1-930643-27-6
   1. Planets--Juvenile literature. [1. Planets. 2. Solar system.] I. Title.

QB602 .K47 2001
523.4--dc21   2001027184

9 8 7 6 5 4 3 2 1

Photo credits
E. Karkoschka (University of Arizona) & NASA/Spacecharts: 22b.
NASA: 29cl.
NASA/Spacecharts: front cover c, 1, 3, 4-5, 10bl, 10br, 11t, 12cl,
13b, 15tl, 15bl, 15r, 16t, 17tr, 18b, 18-19t, 19bl, 19br, 21t, 23tr, 24b, 24-5, 27tr.
Spacecharts: 11b, 12b, 16b, 17bl, 17br, 20bl, 20tr, 23tl, 25tr, 26bl,
27cl, 27bl, 28t, 29tr, 29cr, 30b.
Torleif Svensson/Stockmarket/Corbis: 13t.

Contents

Introducing the planets	4
Moving in circles	6
Little and large	8
Scorching worlds	10
Planet Earth	12
Next-door neighbor	14
Red planet	16
King planet	18
Many moons	20
Ringed planet	22
Rings of ice	24
Distant twins	26
Ice world	28
Mini-planets	29
Probing the planets	30
Useful words	31
Index	32

Introducing the planets

On many nights of the year, some very bright stars appear in the sky. And, month by month, they slowly wander among the other stars. Ancient astronomers called these bright wandering stars **planets**.

But, the wandering stars are not stars at all. Stars are huge, hot balls that give off light. The wandering stars—the planets—are very much smaller and colder. And they give off no light of their own. We see them only because they reflect light from the Sun towards us.

Moving in circles

The planets travel through space with the Sun. They are part of the Sun's family, or **Solar System**. With our eyes and through telescopes, we can see eight planets in the night sky. There is one more planet, which we know very well—the Earth.

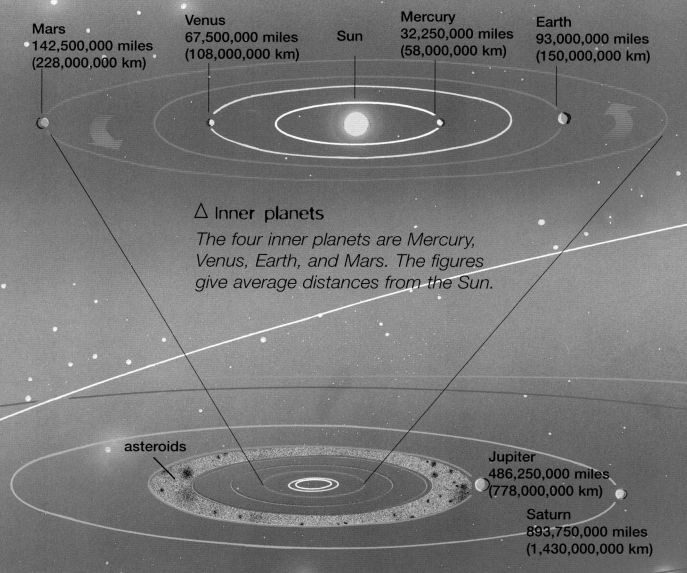

Mars
142,500,000 miles
(228,000,000 km)

Venus
67,500,000 miles
(108,000,000 km)

Sun

Mercury
32,250,000 miles
(58,000,000 km)

Earth
93,000,000 miles
(150,000,000 km)

△ **Inner planets**
The four inner planets are Mercury, Venus, Earth, and Mars. The figures give average distances from the Sun.

asteroids

Jupiter
486,250,000 miles
(778,000,000 km)

Saturn
893,750,000 miles
(1,430,000,000 km)

Round and round

All the planets circle the Sun. The Earth is one of four planets that circle quite close to the Sun, in the inner part of the Solar System. The others are Mercury, Venus, and Mars. After Mars, there is a huge gap before the next planet, Jupiter. In this gap there are thousands of mini-planets, called the **asteroids**.

The planets farther out are very far apart. Saturn is twice as far from the Sun as Jupiter, and Uranus is twice as far away as Saturn. The final two planets, Neptune and Pluto are too far away to be seen with our eyes. The farther a planet is away from the Sun, the longer it takes to circle the Sun. Mercury takes only 88 days to circle the Sun. Pluto takes nearly 250 years.

▽ **Outer planets**
The five outer planets are Jupiter, Saturn, Uranus, Neptune, and Pluto. The figures give average distances from the Sun.

Pluto
3,700,000,000 miles
(5,900,000,000 km)

Uranus
1,794,000,000 miles
(2,870,000,000 km)

Neptune
2,800,000,000 miles
(4,500,000,000 km)

Little and large

To us, the Earth we live on seems huge. But it is one of the smallest of the planets. Earth's neighbors among the planets—Mercury, Venus, and Mars—are even smaller. If you look at the picture, you can see how tiny they are compared with Jupiter, Saturn, Uranus, and Neptune. These four planets are called the giant planets.

Earth and its three neighbors are all solid, rocky bodies—different from the giant planets, which are made up mainly of light gas. We often call them the **gas giants**. Tiny Pluto is in a class by itself. It is made up of a mixture of rock and ice.

Jupiter
88,875 miles
(142,200 km)

Mercury
3,046 miles
(4,874 km)

Venus
7,565 miles
(12,104 km)

Earth
7,973 miles
(12,756 km)

Mars
4,246 miles
(6,794 km)

▷ *The nine planets differ greatly in size, from the very small (Pluto) to the very large (Jupiter). The figures give the diameter of each planet.*

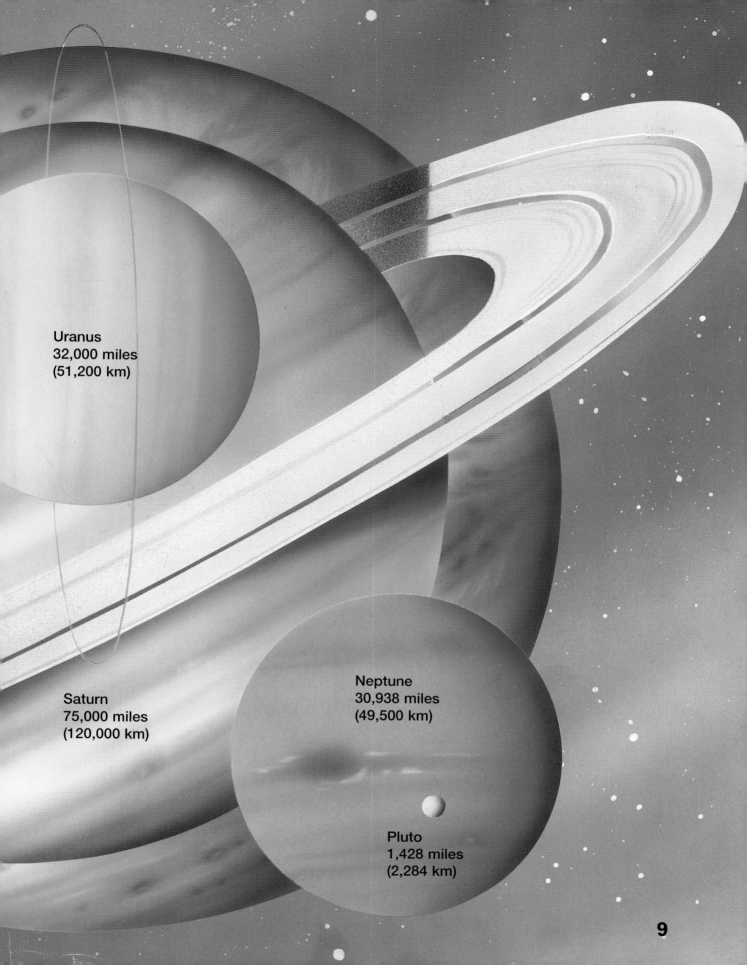

Scorching worlds

Mercury is the planet that is closest to the Sun. It is a small planet, not much bigger than our Moon. And, like the Moon, it is covered in huge **craters**. Mercury is so close to the Sun it gets very hot.

In places, temperatures rise to 840°F (430°C), which is nearly twice as hot as a kitchen oven. Mercury is a difficult planet to spot. It stays close to the Sun in the sky and is usually lost in the Sun's glare.

◁ **Like the Moon**
Craters cover almost all of Mercury, making it look much like the Moon.

▽ **In close-up**
The largest crater here is 95 miles (150 km) across.

Evening star

Venus is an easier planet to find. On many nights it can be seen shining brightly in the west just after sunset. We call it the evening star. Venus is nearly as big as the Earth, and it comes closer to us than any other planet. But it is quite different from the Earth. It has a much thicker **atmosphere** (layer of gases). If you went to Venus, the atmosphere would crush you to death. The thick atmosphere also traps the Sun's heat, making the planet even hotter than Mercury.

△ **Cloudy weather**
Swirling clouds show up in Venus's thick and deadly atmosphere.

▽ **Volcanic activity**
Volcanoes *all over Venus have covered the planet with lava flows.*

Planet Earth

The Earth is home for millions of kinds of living things, from tiny plants to huge animals like whales. There are three reasons why things can live on Earth. It is not too hot and not too cold. It has water on the surface and oxygen in the atmosphere. All living things need warmth, water, and oxygen.

△ **Spaceship Earth**
From space, the Earth looks beautiful. White clouds swirl above the blue seas.

▷ **Fiery fountains**
Fountains of molten lava from an erupting volcano. This shows that Earth is still a very active planet.

Rocky surface

The Earth is made up mainly of rock. But only about a third of the Earth's surface is rocky. This forms the land areas, or continents. The rest of the surface is covered with water, forming the oceans.

△ **Life-giver**
Animals drink at a water hole on the plains of East Africa. Life on Earth depends on water.

On the move

The continents do not stay in the same place all the time. They slowly move around. They sit on rocky islands, which float on a layer of hot, liquid rock. The Earth is not rocky all the way through. In the middle there is a big ball of iron. This turns the Earth into a kind of magnet.

▷ **Drifting continents**
A view from space of Africa (left) and Arabia (right). The two land masses are slowly splitting apart.

Next-door neighbor

The **Moon** lights up the sky on most nights of the year. It is Earth's closest companion in space, only about 240,000 miles (384,000 km) away. The Moon is Earth's only natural satellite. It circles the Earth once a month. During this time, it seems to change shape, becoming bigger or smaller. We call these changes the **phases of the Moon**.

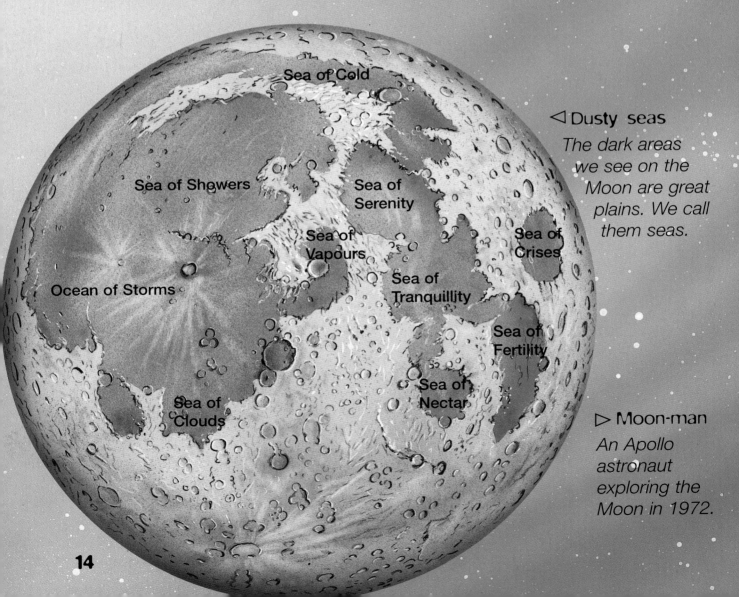

◁ Dusty seas
The dark areas we see on the Moon are great plains. We call them seas.

▷ Moon-man
An Apollo astronaut exploring the Moon in 1972.

◁ Moon rock
The Moon is a rocky body, made up of volcanic rocks like this.

Dead world

The Moon is quite small, only 2,172 miles (3,476 km) across. This is about the same size across as the United States. It does not have any air, atmosphere, or water. Nothing can live on the Moon. It is a dead world.

Like all bodies in space, the Moon has **gravity**, or a pull. And this pull affects the Earth. It tugs at the water of the oceans and makes the tides rise.

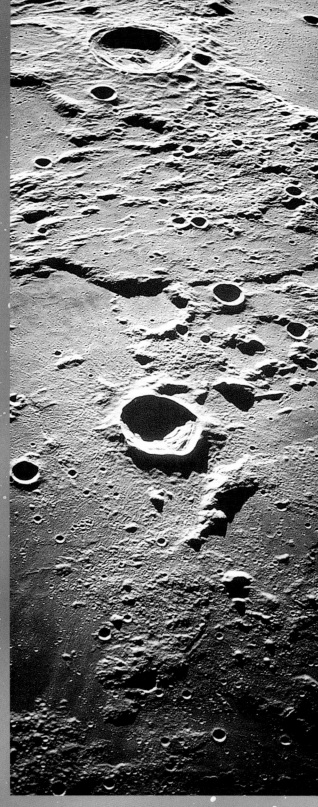

△ Craters galore
Craters large and small pepper the Moon's surface.

Red planet

Mars is quite an easy planet to spot in the night sky. It is often bright and has a reddish glow. That is why we call it the Red planet. Mars is a small world, only about half as big across as the Earth.

Mars is much colder than Earth. Frost often covers the ground. And there are caps (layers) of ice at Mars's north and south poles.

△ **Clouds**
Clouds cling to the slopes of Mars's biggest volcano, Olympus Mons. It is 15½ miles (25 km) high.

▷ **Ice cap**
Ice covers Mars's north pole.

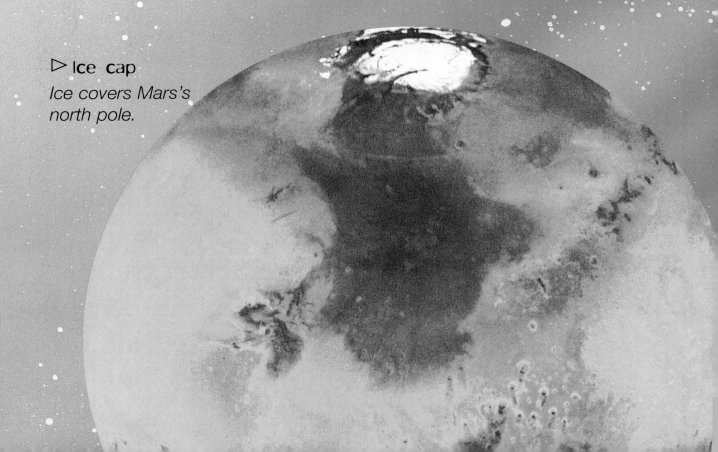

Ancient rivers

Today, there is no liquid water on Mars. But many years ago there were probably rivers and even seas. Mars has many interesting features. It has vast sandy deserts and areas covered with craters. It has towering **volcanoes** and mazes of valleys. The biggest valley is Mariner Valley, which is about 3,125 miles (5,000 km) long. It is much longer, deeper, and wider than America's famous Grand Canyon.

▷ Frost
Frost gathers in a huge basin (top) in this view of a crescent Mars. And clouds cling to Olympus Mons volcano (bottom).

▽ Rocks
Small rocks are scattered all over Mars's surface. They contain a lot of iron, which makes them a reddish color.

King planet

Ancient people named the planet Jupiter after the king of their gods. It is a good name because Jupiter is a king among the planets. It is bigger than all the other planets put together, and could swallow more than 1,000 Earths.

Jupiter shines a brilliant white on many nights of the year. One reason why it shines so bright is that it has a thick, cloudy atmosphere, which reflects sunlight well.

△ **Voyager's view**
The Voyager 2 space probe's view of Jupiter, showing the Great Red Spot (bottom left) and the moon Io.

◁ **Big storm**
A close look at the Great Red Spot. It is an enormous swirling storm that has been raging for centuries.

Colorful atmosphere

In telescopes, Jupiter's atmosphere looks very colorful. It is crossed by pale and dark bands of orange and brown. These bands are clouds rushing round the planet at high speed. In and between the bands, there are great swirls, waves, and colored spots. These are very stormy regions. The biggest storm is a region called the Great Red Spot.

Underneath the atmosphere, Jupiter is completely covered by an ocean. It is not an ocean of water but an ocean of **liquid gas** (hydrogen).

▽ Great balls of fire
In 1994, over 20 bits of a comet (below) hit Jupiter. They caused great fireballs in the atmosphere (left).

Many moons

The Earth has one moon circling round it. But Jupiter has at least 16 moons. We can see the four biggest in binoculars. They are Io, Europa, Ganymede, and Callisto.

Ganymede is the largest moon in the Solar System. It is bigger than the planet Mercury. Callisto is slightly smaller, but still bigger than Mercury. Io is about the same size as our Moon, while Europa is slightly smaller.

△ Ganymede
Diameter 3,297 miles (5,276 km)

◁ Io
Diameter 2,270 miles (3,632 km)

Rock and ice

Ganymede, Callisto, and Europa are made up of rock and ice. Ganymede has a dark surface with many white spots. These spots show where **meteorites** have hit the surface and thrown out fresh white ice. Callisto is darker and has many more craters. Europa is pale and very smooth. It is covered with cracked ice.

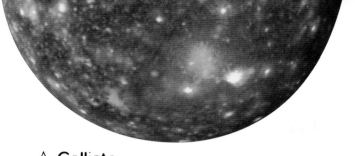

△ **Callisto**
Diameter 3,012 miles (4,820 km)

Rivers of sulfur

Io is completely different. It is vividly colored yellow and orange. And volcanoes are erupting in places. They don't pour out liquid rock, but rivers of yellow sulfur.

△ **Io's eruptions**
Volcanoes erupt on Io. Molten sulfur pours onto the surface. Sulfur fumes spurt into the sky.

Ringed planet

Saturn is the second biggest planet, just smaller than Jupiter. But it never shines very brightly in the night sky because it is so far away—twice as far as Jupiter. In a telescope, Saturn looks different from all the other planets. This is because it has **rings** circling it. Saturn is similar to Jupiter. It is a gas giant, with bands of clouds in its atmosphere. Underneath the atmosphere there is a vast ocean of liquid gas.

◁ **Stormy weather**
Bands of clouds and storms show up in Saturn's atmosphere.

△ **Thin rings**
Saturn's rings are thin and sometimes almost disappear from view, as they did in 1995 (top).

△ **Cratered moons**
Saturn and five of its larger moons. Like the other moons, Dione is pitted with craters.

Many Moons

Like Jupiter, Saturn has many moons circling it—at least 18. The largest ones seem to be made up of rock and ice.

Biggest of all is Titan, about 3,212 miles (5,140 km) across.

Titan is the second largest moon in the Solar System. It is unusual because it has a thick atmosphere. Astronomers think that it may rain and snow liquid and frozen gas on Titan.

Rings of ice

Saturn has three main rings, which we can see from Earth. It also has several narrow faint rings that we can't see. All these rings are made up of millions upon millions of bits of ice, whizzing round at high speed. Some lumps are as big as boulders, others as small as pebbles.

▷ **Ringside**
What the rings would look like close-up.

△ **Ringlets**
Thousands of ringlets form the rings.

△ Different particles
This picture shows up differences between the particles that make up the ringlets.

Mysterious rings

Space probes have taken close-up pictures of Saturn's rings. The three rings we see from Earth are actually made up of thousands of ringlets. We don't know where the rings began. They may be the remains of a moon, torn to pieces by Saturn's powerful gravity, or pull.

Distant twins

For centuries, people thought that Saturn was the most distant planet. But, in 1781, an English astronomer named William Herschel discovered another one, twice as far away as Saturn. It was called Uranus.

In 1846, astronomers found a second new planet, twice as far away as Uranus. They called it Neptune. Astronomers realized that the Solar System was much bigger than they had thought.

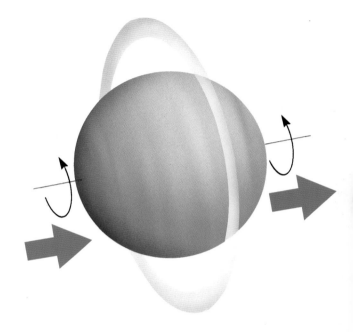

△ **Topsy-turvy**
Uranus spins on its side as it circles in orbit around the Sun.

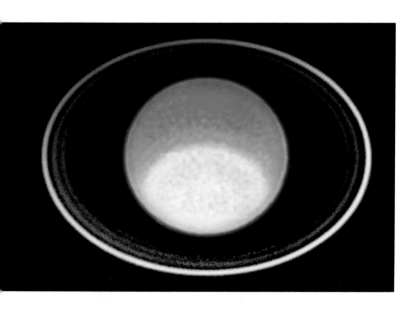

◁ **Rings round Uranus**
Uranus has 11 faint rings circling round it.

▷ **Amazing Miranda**
Uranus's moon Miranda has the strangest markings on its surface.

▷ **Neptune's atmosphere**
Flecks of clouds and dark spots appear in Neptune's atmosphere. The big dark spot is a stormy region.

▽ **Triton's geysers**
Dark fumes erupt from geysers on Neptune's largest moon, Triton.

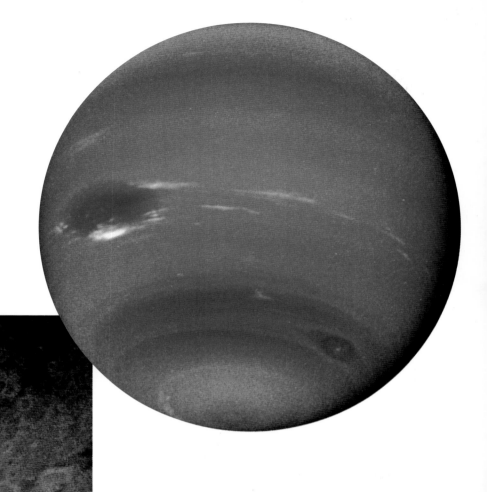

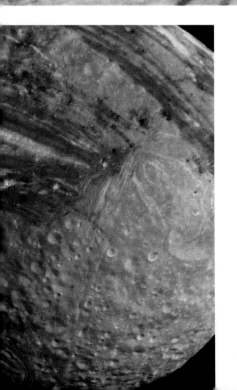

Vast oceans

Uranus and Neptune are nearly identical in size—about four times bigger across than the Earth. They both have a deep atmosphere and, underneath, an ocean containing water and liquid gases. They are both bluish in color.

Uranus has more than 20 moons. It also has some faint rings circling it. Neptune has faint rings, too, but only eight moons. One, Triton, is nearly as big as our Moon.

Ice world

After astronomers found Neptune, they carried on looking for new planets. But it was not until 1930 that they found another one. It was called Pluto. Pluto is the smallest planet, made up of rock and water ice. Its surface is covered with frozen gases. Gases freeze into solid ice when they get very cold. And Pluto is very cold (-382°F/-230°C).

Pluto has one moon, called Charon. Surprisingly, it is half the size of Pluto. And it always stays in the same spot in Pluto's sky. No other moons do this.

△ Half-size
Pluto and its moon Charon, which is half its size.

▽ Moonrise
Charon appears huge from the deep-frozen surface of Pluto.

Mini-planets

The planets formed millions of years ago when lumps of rock and other materials smashed into one another and stuck together. Some lumps were left over. We find many of them today circling the Sun between Mars and Jupiter. We call these lumps the asteroids.

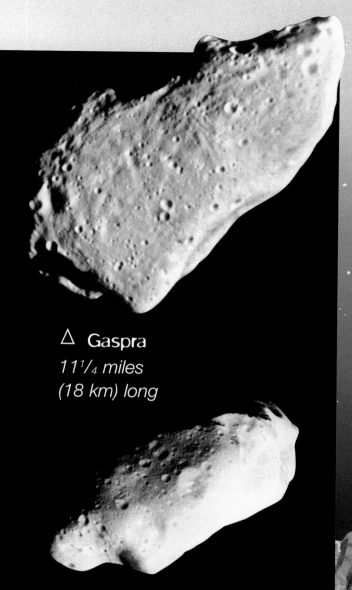

△ Gaspra
11 1/4 miles
(18 km) long

▷ Eros
15 miles
(24 km) long

△ Ida
35 miles (56 km) long

Asteroid hit

There are thousands of asteroids. The largest is Ceres, about 625 miles (1,000 km) across. But most are much smaller. Most asteroids circle the Sun in a band, called the asteroid belt. But some wander among the planets. A few sometimes stray near the Earth. Millions of years ago, asteroids may have hit the Earth. The dinosaurs might have been wiped out by an asteroid hit.

Probing the planets

Years ago, we could only study the planets from afar with our eyes and through telescopes. But today, we can study the planets close-up using spacecraft we call **probes**.

Space probes are launched by rocket or space shuttles and may take years to reach their target planet. They carry cameras and other instruments to record data—such as a planet's weather. Since the 1960s, space probes have explored all the planets except Pluto. The probe *Mariner 10* (1974) gave us our first pictures of Mercury, and *Magellan* (1990) revealed the great lava plains of Venus. *Cassini* will reach Saturn in 2004. But the most successful probe is *Voyager 2* (1977). It has visited Jupiter, Saturn, Uranus and Neptune.

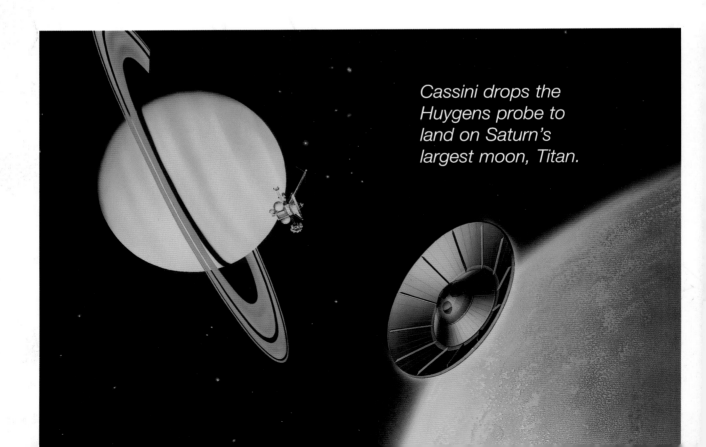

Cassini drops the Huygens probe to land on Saturn's largest moon, Titan.

Useful words

asteroids Lumps of rock that circle the Sun between Mars and Jupiter.

atmosphere The layer of gases around a planet.

crater A hole in the surface of a planet or moon, made by a meteorite hitting it.

gas giant One of the four biggest planets, Jupiter, Saturn, Uranus, or Neptune.

gravity The pull every lump of matter has on every other lump.

liquid gas A gas that has turned into liquid.

meteorite A lump of rock or metal from outer space.

moon A smaller body that circles around a planet.

phases of the Moon The changing shapes of the Moon during the month.

planet A large body that circles the Sun.

probe A spacecraft sent to explore distant planets.

ring A ring of particles circling round a planet.

Solar System The Sun's family, including the planets, their moons, and comets.

volcano A mountain or hole in a planet's surface that pours out molten rock or other material.

Index

asteroids 7, 29, 31
atmospheres 11, 19, 22, 27, 31

Callisto 20, 21
Cassini probe 30
Ceres 29
Charon 28
clouds 11, 16, 17, 22
craters 10, 15, 23, 31

Dione 23

Earth 6, 8, 12–13, 16
Enceladus 23
Eros 29
Europa 20, 21
evening star 11

fireballs 19

Ganymede 20–21
gas giants 8, 31
Gaspra 29
gravity 15, 31
Great Red Spot 18, 19

ice 16, 24, 28
Ida 29
Io 20, 21

Jupiter 6, 8–9, 18–21
liquid gas 19, 22, 31

Magellan probe 30
Mariner 10 probe 30
Mars 6, 8, 16–17

Mercury 6, 8, 10
meteorites 21, 31
Mimas 23
Miranda 26–27
Moon (Earth's) 10, 14–15, 26–27
moons 20–21, 23, 26–27, 28, 31

Neptune 7, 8-9, 26-27

oceans 19, 27

phases of the Moon 14, 31
Pluto 7, 8–9, 28
probes 30, 31

Red planet 16
Rhea 23
ringlets 24, 25
rings 22–25, 26, 31
rocks 13, 15, 17, 29

Saturn 6, 8–9, 22–25
sizes of planets 8–9
Solar System 6, 31
storms 18, 19, 22

Tethys 23
Titan 23
Triton 27

Uranus 7, 8–9, 26–27

Venus 6, 8, 11
volcanoes 11, 12, 17, 21, 31
Voyager 2 probe 30

SOUTH KELOWNA
RESOURCE CENTRE